QA29.P775 A3 1987
Polya, George,
3 4369 00022803 6

D1269355

George Pólya

The Pólya Picture Album: Encounters of a Mathematician

Edited by
G. L. Alexanderson

Birkhäuser
Boston · Basel

Gerald L. Alexanderson
Department of Mathematics
Santa Clara University
Santa Clara, CA 95053, USA

Library of Congress Cataloging-in-Publication Data
Pólya, George, 1887–1985
The Pólya picture album.
1. Pólya, George, 1887–1985. 2. Mathematicians–
Hungary–Biography. 3. Mathematicians–Portraits.
I. Alexanderson, Gerald L. II. Title.
QA29.P775A3 1987 510'.92'2 87-15063
ISBN 3-7643-3352-9

CIP-Kurztitelaufnahme der Deutschen Bibliothek
The *Pólya picture album: encounters of a mathematician* /
George Pólya. Ed. by G. L. Alexanderson.
Basel ; Boston: Birkhäuser, 1987.
ISBN 3-7643-3352-9
NE: Pólya, George [Mitarb.]; Alexanderson, Gerald L. [Hrsg.]

Book design by Albert Gomm, Basel
Typeset, printed and bound in Germany
9 8 7 6 5 4 3 2 1
ISBN 0-8176-3352-9
ISBN 3-7643-3352-9

Contents

Introduction

Over the years mathematicians visiting Stanford sometimes enjoyed guided tours through the Pólya photograph album led by Professor Pólya himself. He identified pictures and passed along personal reminiscences and stories about individuals and meetings represented there. The album spans the greater part of the twentieth century and, on occasion, reaches back to the nineteenth. A few years ago, at the urging of Donald J. Albers, then editor of the *Two-Year College Mathematics Journal* (now the *College Mathematics Journal*), Pólya and I prepared a small selection of pictures with some descriptive material and anecdotes that usually accompanied tours through the album. This appeared in the September 1983 issue of the *Journal.*

Encouraged by the response to this we put together a more extensive selection of pictures from the collection so that a wider audience can enjoy this encounter with some of the mathematical giants of this century. The pictures derive from a variety of sources, but many of the most interesting were taken by Stella Weber Pólya at mathematical meetings in Europe in the 1920's and 30's and later in Palo Alto. She not only provided most of the photographs but maintained the collection as well. To her is due most of the credit for the present effort.

Other photographs have been contributed to the collection by friends. They were given by so many and, in many cases, a number of years ago, it is now impossible to identify in all cases the sources. We are grateful to all these friends, but we are especially grateful to Professor Morris Marden of the University of Wisconsin, Milwaukee, for pictures contributed to the collection in the 1930's, and to Professor Marion Walter of the University of Oregon, who contributed a number of pictures taken in the 1950's.

The album does not cover the full range of mathematics.

Instead it emphasizes mainly those mathematicians who shared with the owner mathematical interests, primarily in the fields of real and complex analysis, probability, number theory, and combinatorics. Pólya's interests spanned a broader range of mathematical fields than most, so many important mathematicians of the era are here. But it is still a personal collection, one man's view of the mathematical community in the twentieth century.

The remarks accompanying the pictures are largely taken from tapes of conversations with visitors, so they approximate as closely as possible Pólya's words when showing the pictures.

Unfortunately, before this project could be completed, Pólya died. Most of the manuscript had been reviewed by him before his death, but later additions and alterations have been reviewed by his wife, Stella Pólya.

We are grateful to many of our friends who have helped us track down dates and specific information about the subjects and who have read drafts of the manuscript. Especially helpful were Harold Bacon, Karel de Bouvère, Dave Logothetti, Lee Lorch, Morris Marden, István Mócsy, Jean Pedersen, Klaus Peters, M. M. Schiffer, I. J. Schoenberg, Marion Walter, and John E. Wetzel. We are also grateful to Paul R. Halmos, Lester H. Lange, William Schulz, and the archivists of the University of Göttingen and Purdue University for help on specific elusive details. The errors are, of course, mine.

For the skillful work on the photographs we are grateful to Dave Jackson. And for all the other help in putting together this manuscript, we wish to thank Mary Jackson, and Yvonne Pasos Sullivan, as well as the very helpful staff at Birkhäuser, especially George Adelman, Carol Munroe and Gaby Seiler.

June, 1986
G. L. Alexanderson
Santa Clara, California

George Pólya: A Biographical Sketch

by G. L. Alexanderson

George Pólya has a special place in mathematics. Admired by research mathematicians for his original and lasting contributions to mathematics itself, he was revered by his students and loved by his colleagues. He was a pioneer respected by all teachers of mathematics for his seminal work in heuristics and problem solving. And he had a long career as researcher, scholar, and teacher. During this long career he knew many of the important mathematicians of his era and collaborated with an impressively large number of them.

He was born in Budapest on December 13, 1887, the son of Jakob and Anna (Deutsch) Pólya. Early in life he was urged by his mother to take up his father's profession, the law. Eugene (Jenő), his older brother by eleven years, became an eminent surgeon in whose honor a now standard form of stomach surgery is named. As a gymnasium student, George Pólya was not particularly attracted by mathematics but had outstanding teachers in geography, Latin, and Hungarian. Of his three mathematics teachers at the gymnasium, he is quoted as saying that "two were despicable and one was good." [16]

During the years when Pólya was a student, many of his contemporaries were participating in a prestigious problem contest in mathematics, the famous Eötvös Competition, named for the illustrious Hungarian physicist, Loránd Eötvös. This contest is often credited in part for the extraordinary number of first-class mathematicians who emerged in Hungary in the latter part of the 19th and first part of the 20th centuries. Pólya, when asked to take the contest, failed even to turn in his paper.

He dutifully began his work at the University of Budapest by studying law, but this lasted only for one semester. He found it boring. He then studied languages and literature for two years, after which he passed an examination that resulted in a lifetime teaching certificate allowing him to teach Latin and Hungarian in the lower grades of the gymnasium. This certificate he never used.

Fond of philosophy and of literature, particularly the poetry of Heinrich Heine (some of which he translated into Hungarian while still in the gymnasium), he was also attracted to physics at that time of his life. A philosophy professor with unusual insight convinced him that the study of mathematics and physics would help in his understanding of philosophy, so he eventually came to the serious study of mathematics. This led to the now well-known Pólya quote: "I thought I am not good enough for physics and I am too good for philosophy. Mathematics is in between." [1; p. 248]

He was influenced by many of his professors at the University of Budapest – his physics professor was Eötvös, mentioned earlier. But the one to influence him most was the mathematician Lipót Fejér. Fejér had an appealing style and personality that drew a number of very talented students into his mathematical circle, many of whom are represented in Pólya's photograph collection: Mihály Fekete, Ottó Szász, Gábor Szegő, and, much later, Paul Erdős. Pólya's reminiscences of Fejér appear in [9].

Pólya spent the 1910–11 academic year at the University of Vienna, and returned to Budapest to receive his doctorate in 1912. He then went to the University of Göttingen in 1912–1914 where he encountered Felix Klein, David Hilbert, Carl Runge, Edmund Landau, and other eminent Göttingen professors. Even the list of Privatdozents was impressive: Hermann Weyl, Erich Hecke, Richard Courant, and Otto Toeplitz. So even in light of the illustrious mathematical and scientific tradition at Göttingen through the 19th century – Gauss, Dedekind, and Riemann had been there, and Minkowski had died only three years before Pólya's arrival – the early part of the 20th century surely stands out as the mathematical golden age in Göttingen.

In the spring of 1914 he went to the University of Paris, where he met Émile Picard and Jacques Hadamard, who would influence his later work. In the fall of that year, at the invitation of Adolf Hurwitz, he took his first teaching position at the

Eidgenössische Technische Hochschule (ETH, the Swiss Federal Institute of Technology) in Zürich, where he was to stay until 1940 and to which he returned for frequent visits later. During the First World War he was initially rejected by the Hungarian Army because of the aftereffects of infection in his leg resulting from a soccer injury. Later, when the need for soldiers became too great, the Hungarian Army again asked him to report, but by that time he had fallen under the influence of Bertrand Russell, the English mathematician-philosopher, who was an outspoken pacifist. Pólya refused to serve and was therefore unable to return to Hungary for many years.

He became a Swiss citizen and in 1918 married Stella Vera Weber, daughter of a professor of physics at the University of Neuchâtel. Because Mrs. Pólya grew up in French-speaking Switzerland, French was the language spoken in the home, though Pólya was, of course, Hungarian, and they lived in German-speaking Switzerland. Living with these three languages (English came later), in which he felt equally at home, fitted nicely with his interest in languages. He wrote mathematical papers in Hungarian, French, German, Italian, English, and Danish. Beyond these he had studied Latin and Greek in school and could read, with varying degrees of ease, several other languages as well.

At the ETH he rose through the ranks to Professor in 1928 and enjoyed the association with outstanding colleagues, among them Adolf Hurwitz, the colleague who he felt influenced him most. Arthur Hirsch was Department Head in the early years. Later his colleagues at the ETH included Hermann Weyl, Michel Plancherel, and Heinz Hopf. In 1924 he spent a year in England as the first international Rockefeller Fellow, initially at New College, Oxford, then at Trinity College, Cambridge, in order to work with G. H. Hardy. It was Hardy who had recommended Pólya when asked by the Rockefeller administrators to suggest a recipient for the grant. During that time work was started on the authoritative book *Inequalities,* written with Hardy and Littlewood and published by the Cambridge University Press in 1934.

Earlier Pólya had developed a close collaboration with another Hungarian mathematician, Gábor Szegő, who had received his

Ph. D. from the University of Vienna in 1918 and who was appointed Privatdozent in Berlin in 1921. Pólya wrote with Szegő what was probably the most important single work of either one, the now classic *Aufgaben und Lehrsätze aus der Analysis,* published in two volumes by Springer-Verlag in 1925. After 60 years, this work is still cited regularly and is one of the most important sources for problems in analysis. The organization was original: the problems were put together not according to the topic, but according to the method of solution. The first volume was translated into English in 1972 and the second in 1976; the paperback edition appeared in 1976 at which time the problem sets were revised and expanded.
This famous work is one to which the word 'masterpiece' is often applied – and with justification. The collaboration with Szegő continued with joint papers and another book, *Isoperimetric Inequalities in Mathematical Physics,* published as part of the Annals of Mathematics Studies by the Princeton University Press in 1951.
During the 1930's Pólya worked closely with Gaston Julia on a series of problems, and this collaboration meant fairly frequent trips to the Institut Henri Poincaré in Paris. Then in 1933 Pólya was again selected for a Rockefeller grant, this time to visit Princeton University. Though there were no mathematicians at Princeton with whom he worked closely, he discussed questions with Oswald Veblen and met various other mathematicians on the East Coast. During the summer Pólya visited Stanford University at the invitation of H. F. Blichfeldt. This was the Pólyas' first trip to the United States and they both decided they liked California and, in particular, Stanford very much.
In 1940 when it became quite clear that Europe was being enveloped by the chaos of the Second World War, the Pólyas decided to leave Switzerland. They travelled by way of Lisbon to the United States. With so many academic emigrés leaving Europe at that time, many desperate to escape the Nazis in Germany, Poland, and other countries overrun by Hitler's troops, suitable positions were not easily available in the United States. After two years at Brown University and a short stay at Smith College, Pólya received an appointment at Stanford University where Szegő was, by that time, Department

Head. In January of 1942 Pólya stayed in the East, while Mrs. Pólya travelled by train to California to buy their house in Palo Alto.

The Department at Stanford had some distinguished faculty before the War, notably H. F. Blichfeldt, J. V. Uspensky, and W. A. Manning. Blichfeldt had extensive contacts in Europe, so there had always been important visitors like Harald Bohr, Harold Davenport, Alfred Errera, and Edmund Landau. But with the arrival of many European mathematicians, after the Second World War Stanford could boast a number of first-class mathematicians in addition to Szegő and Pólya. Stefan Bergman, Charles Loewner, and Menahem Schiffer come immediately to mind.

Prior to coming to America, Pólya started a manuscript for a book on problem solving that became his most popular book, *How To Solve It.* Hardy thought that the title would be successful in the United States because Americans like "How To" books. This title, originally published by the Princeton University Press, was issued in paperback by Doubleday and has by now sold well over one million copies. It has also been translated into at least 17 other languages, almost certainly a record for a modern mathematics book. Following the enormous success of *How To Solve It,* Pólya wrote the very beautiful two volume set, *Mathematics and Plausible Reasoning,* again illustrating some of the heuristic principles set out earlier in *How To Solve It* and some of his articles. Written at a considerably more sophisticated mathematical level than *How To Solve It,* this set was published in 1954. It was followed by a more elementary set, *Mathematical Discovery,* again in two volumes, the first in 1962, the second in 1965. (It was reissued in one volume, paperback, in 1981.) These established him as the foremost advocate of problem solving and heuristics in his generation. Though he had distinguished antecedents who had also written about heuristics and the psychology of problem solving, from Descartes to Hadamard, Pólya nevertheless is the father of the current trend toward an emphasis on problem solving in mathematics teaching.

After what was termed his retirement from active service at Stanford in 1953, he continued to teach and write. He became more and more interested in the teaching of teachers and taught in a series of teacher institutes supported by General Electric, Shell, and the National Science Foundation during the 50's, 60's and 70's at Stanford and, in the summer of 1964, at a Stanford program outside Geneva, Switzerland. It was during this time that he was awarded a number of honorary degrees – by the University of Alberta, the University of Wisconsin at Milwaukee, the University of Waterloo, and the Swiss Federal Institute of Technology – and memberships in several national academies. He was corresponding member of the very prestigious Académie des Sciences, Paris, as well as the American Academy of Arts and Sciences, the Hungarian Academy of Sciences, the National Academy of Sciences of the United States, and the Académie Internationale de Philosophie des Sciences, Bruxelles. He was also an honorary member of the London Mathematical Society, the Swiss Mathematical Society, the New York Academy of Sciences, and the Council of the Société Mathématique de France. In 1963 he was given the Award for Distinguished Service to Mathematics by the Mathematical Association of America, and in 1968 he was given the Blue Ribbon by the Educational Film Library Association for his film 'Let Us Teach Guessing.'

In addition to the books already cited he wrote a text *Complex Variables* with Gordon Latta, and several other books and monographs: *The Stanford Mathematics Problem Book* (with Jeremy Kilpatrick), 1974; *Mathematical Methods in Science* (edited by Leon Bowden), 1963, 1977; and *Notes on Introductory Combinatorics* (lectures by Pólya and Robert Tarjan, notes by Donald Woods), 1984. His influential paper on combinatorics [8] will appear in English translation with commentary by R. C. Read as a monograph entitled *Combinatorial Enumeration of Groups, Graphs, and Chemical Compounds*. This is scheduled for publication in 1987 by Springer-Verlag. He wrote or coauthored over 250 papers and these have been collected together in four volumes published by the MIT Press [7].

One of the remarkable aspects of Pólya's career was his mathematical breadth, mirrored in part by the range of interests of mathematicians with whom he coauthored mathematical papers: Issai Schur, Mihály Fekete, Adolf Hurwitz, Rudolf Fueter, G. H. Hardy, J. E. Littlewood, A. E. Ingham, Michel Plancherel, Norbert Wiener, Harold Davenport, Menahem Schiffer, and I. J. Schoenberg, among others. When asked why he had worked on problems from such seemingly diverse branches of mathematics as real and complex analysis, calculus of variations, mathematical physics, probability, geometry, number theory, combinatorics, and graph theory, Pólya responded: "...I was influenced by my interest in discovery. I looked at a few questions just to find out how you handle this kind of question." [1; p. 247]

Apparently Hardy disapproved, to some extent, of his moving too quickly at times from one problem to another. Pólya related a story about this. "I once had an idea of which [Hardy] approved. But afterwards I did not work sufficiently hard to carry out that idea, and Hardy disapproved. He did not tell me so, of course, yet it came out when he visited a zoological garden in Sweden with Marcel Riesz. In a cage there was a bear. The cage had a gate, and on the gate there was a lock. The bear sniffed at the lock, hit it with his paw, then he growled a little, turned around and walked away. 'He is like Pólya', said Hardy. 'He has excellent ideas, but does not carry them out.'" [9]

Pólya nevertheless made remarkable contributions in divers branches of mathematics. His doctoral dissertation was on "Some questions of the calculus of probability, and some definite integrals associated with it." (There was no one at Budapest interested in probability, so he wrote this without the help of an advisor.) This started a series of investigations in probability that led to some of his best known work. Early papers explored some aspects of geometric probability, questions of probabilities of events given in terms of random variables that are lines, or planes, or other geometrical objects. Feller [7; IV, p. 608] claims that Pólya was the first to use in written work the term 'Central Limit Theorem' to describe the normal limit law in probability, although there seems to be evidence that the phrase was in use earlier.

Pólya worked further in so-called characteristic functions in probability theory, for which there is a "Pólya criterion". And one example of his work is known to every probability student, the Pólya urn scheme where one considers an urn containing r red balls and b black ones, say. When a ball is drawn at random, it is replaced and, in addition, c balls of the same color drawn are added to the urn. A limit distribution exists and the model is often used to describe contagion, where an event affects the probability of future events. An offshoot of this model is the so-called Pólya distribution.

One of Pólya's most interesting contributions to probability was his 1921 paper on random walks [11], in which, incidentally, the phrase 'random walk' was used for the first time. Imagine an infinite plane with lines equally spaced meeting at right angles. It would look like a street map of a town that is boringly regular – all blocks are the same size, all streets meet at right angles, and the town goes on and on. Now imagine a man standing at a certain street corner. He has four different directions he can go: north, south, east, or west. Suppose the choice of each direction is equiprobable, with probability equal to one-fourth. Suppose further that the man faces the same decision after he has walked the first block, the second block, and so on, and the probabilities of going in any of the four directions are again one-fourth at any intersection. Then this is called a simple random walk in two dimensions. A one-dimensional random walk would involve starting at a point corresponding to an integer on the real line and having to move forward or backward on the line a unit distance, with probability one-half of making either choice. A possible interpretation of this would be the fate of a gambler betting on heads or tails in repeated tosses of a fair coin. After so many tosses, or so many moves along the line, the final position of the moving point would correspond to the total take (or total loss) of the gambler.

Now consider a lattice in d dimensions with mutually perpendicular lines connecting points with integer coordinates. At each 'vertex', the intersection of d lines, there would be $2d$ edges from which to choose, with the probability of choosing any one of the edges $1/(2d)$. In his 1921 paper, Pólya proved a striking theorem. In a one- or two-dimensional lattice, the wandering point must return to its starting point given sufficient time. But this need not happen in any higher dimension.

In two dimensions, then, in Pólya's own description, all roads really do lead to Rome!
At the New York World's Fair in 1964, IBM had in its pavilion a display demonstrating random walk. The principal contributors to the subject were listed: Albert Einstein, Enrico Fermi, Norbert Wiener, Alexander Kolmogorov, John von Neumann, George Pólya, and Stanislaw Ulam. Pólya remarked, when Stefan Bergman sent a picture of the display to him, that he liked the company he was in.
Yet, striking though such results in probability may be, Pólya's most profound and difficult work is probably in complex function theory, particularly in the study of entire functions, single valued functions having no singularities in the finite part of the plane. Unfortunately, since most of these results are rather specialized it is difficult to describe their significance without a good deal of technical language. But it is an indication of the level of Pólya's contribution in this field that part of this technical terminology includes terms such as 'Pólya peaks', 'the Pólya representation', and 'the Pólya gap theorem'. The list of others working in this field includes some of the most powerful mathematicians of the day: Gaston Julia and Jacques Hadamard in France, Rolf Nevanlinna in Finland, G. H. Hardy in England, Oskar Perron and Ludwig Bieberbach in Germany, Mihály Fekete in Hungary, and, among later mathematicians, Atle Selberg in Norway, Alfréd Rényi of Hungary, Mary Cartwright in England, Alexander Ostrowski of Russia, and Albert Pfluger of Switzerland, Pólya's Ph. D. student and his successor at the ETH.
The Pólya-Schur functions, investigated in a joint paper with Issai Schur in 1914, have arisen in a number of subsequent investigations, including the work of I. J. Schoenberg on approximation by spline functions. In 1957 Pólya and Schoenberg stated a conjecture concerning power series that map the unit circle into convex domains: the Hadamard product of two power series with this property is again a power series having this property. This became known as the Pólya-Schoenberg conjecture. After 15 years of efforts by several mathematicians, it was finally established in 1973 by St. Ruscheweyh, of Würzburg, and T. Sheil-Small, of York [15]. In connection with a

paper by Pólya from 1915, Schoenberg in 1947 solved a moment problem that led him to introduce certain frequency functions he called Pólya Frequency Functions. They have such diverse properties that N. G. de Bruijn proposed that they should be called 'Polyamials.'
Some of Pólya's most interesting contributions to the theory of functions concern the zeros of functions and, in particular, the fate of the zeros upon differentiation. Work in this area is often close to the subject of the celebrated Riemann Hypothesis, to which so many of the best mathematicians of the early part of the 20th century devoted so much of their time. There are rumors from time to time that this question has finally been settled, but at present the problem is apparently still open. This famous problem has, however, prompted some of the deepest mathematics of our time and, as a bonus, has stimulated the development of some of the most entertaining mathematical lore. For some amusing accounts see [9].
One paper of Pólya's in 1926 [6] apparently came close to taking care of the Riemann Hypothesis. Though it failed, it did lead to important developments in statistical mechanics. Another paper of Pólya's, in 1919 [13], contained a conjecture, since called the Pólya Conjecture, that would have implied the Riemann Hypothesis, had it been true. The Conjecture states that for each $x > 1$ there are at least as many integers $\leq x$ having an odd number of prime factors (not distinct) as there are numbers with an even number of prime factors. It is, in a sense, plausible since the primes themselves have an odd number of prime factors and one reaches in the sequence of integers all the primes before one can reach the composite numbers of which those primes are factors. This conjecture was widely felt to be true until 1958 when C. B. Haselgrove proved there are infinitely many counterexamples [5], though he did not produce any. R. S. Lehman found a counterexample in 1962: 906180359. Pólya's contributions to number theory are largely in the area of analytic number theory, various asymptotic formulas, questions of kth power residues and nonresidues, and estimates on the size of the largest prime factors of certain polynomials.
In the late 1940's he wrote a number of papers on differential

equations and a series of several articles on mathematical physics. Some of this material appeared later in the book with Szegő, *Isoperimetric Inequalities in Mathematical Physics,* mentioned earlier. A series on isoperimetric problems, vibrating membranes, and their eigenvalues continued until 1960, with coauthors M. M. Schiffer, L. E. Payne, and H. F. Weinberger.

The oldest isoperimetric problem goes back to antiquity; it is the so-called problem of Dido. It asks for a plane region of given area with least perimeter or, equivalently, for the largest area with given perimeter. A number of similar problems arise with the development of mathematical physics. One of the best known was formulated by Lord Rayleigh: What drum with given membrane area vibrates with minimum frequency? In this problem as well as in that of Dido the obvious shape of the region has to be a circle. But the proof in each case is not easy. One of the most ingenious and intuitive solutions of the problem of Dido is due to the Swiss geometer, Jakob Steiner. He developed a method called 'symmetrization' which deforms a given region into one that is symmetric in a given direction, and has the same area but a smaller perimeter. From this insight it is easy to show that the most symmetric figure, that is, the circle, solves the problem of Dido. Pólya showed that the same method works for many analogous problems in geometry and mathematical physics. He gave, in particular, the most elegant solution of the Rayleigh problem.

Early in his career (1913) Pólya made an interesting contribution to geometry [12]. He described a Peano curve, a curve that passes through every point of a region, that has the additional property that it passes at most three times through any of the points of the region. It is well-known that such curves must have points that are at least triple points, but that such curves need not have points of higher multiplicity is interesting. A further contribution to geometry was his famous paper of 1924 in which he described the 17 symmetries of the plane [10]. The Dutch artist M. C. Escher studied this paper when it was called to his attention by his brother. Soon after that, additional symmetries began to appear in Escher's etchings and prints. Pólya and Escher even corresponded prior to the Second World

War, but most of the correspondence seems, unfortunately, to have been lost.

Pólya's interest in symmetry emerged again in 1935 in a series of papers on isomers in chemistry, culminating in his monumental paper of 1937 on groups, graphs, and molecular structures [8]. One of the high points in the history of combinatorics, this paper generalized Burnside's lemma (Frobenius's theorem) to count essentially different patterns, patterns that could not be changed into each other by geometrical transformations such as rotation in space. A paper by the American, J. H. Redfield, had apparently anticipated this work as early as 1927 [14], but the paper had been completely ignored and, as with so many discoveries in history, Redfield's contribution – and it is sad for Redfield – is a footnote in the history of combinatorics. Pólya's work was accessible and comprehensive, and the principal theorem is now cited in any combinatorics text as the 'Pólya Enumeration Theorem'. It provides a powerful and subtle technique (that at the same time is accessible to those with only an elementary mathematical background) for counting graphs, geometrical patterns, and, not surprisingly, chemical compounds.

It is partly in recognition of this work in combinatorics, which has resulted in numerous applications, that the Society for Industrial and Applied Mathematics established their Pólya Prize in Combinatorial Theory and its Applications. Other awards named for Pólya are the Pólya Prize for Expository Writing in the *College Mathematics Journal,* given by the Mathematical Association of America, and the Pólya Prizes in the Problem Solving Competitions in *The Mathematics Student,* given by the National Council of Teachers of Mathematics (1978–80).

Finally, let us look at remarks by two contemporary mathematicians who have written widely on extensions of the work of Pólya on combinatorics and graph theory. In his retirement address from the Technological University Eindhoven, N. G. de Bruijn said [3]: "A mathematician who possibly more than anyone else has given direction to my own mathematical activity is G. Pólya. All his work radiates the cheerfulness of his personality. Wonderful taste, crystal clear methodology, simple means, powerful results. If I would be asked whether I could name just one mathematician who I would have liked to be myself, I have my answer ready at once: Pólya." And in his introduction to a special issue of the *Journal of Graph Theory* that honored Pólya's 90th birthday, Frank Harary wrote [4]: "I must mention that George Pólya is not only a distinguished gentleman but a most kind and gentle man: his ebullient enthusiasm, the twinkle in his eye, his tremendous curiosity, his generosity with his time, his spry energetic walk, his warm genuine friendliness, his welcoming visitors into his home and showing them his pictures of great mathematicians he has known – these are all components of his happy personality. As a mathematician, his depth, speed, brilliance, versatility, power, and universality are all inspiring. Would that there were a way of teaching and learning these traits!"

Pólya died in Palo Alto on September 7, 1985.

For a bibliography of Pólya's papers, see [7], for a detailed account of his books, including information on translations, see [2].

References

[1] Albers, Donald J. and G. L. Alexanderson. **Mathematical People: Profiles and Interviews.** Boston: Birkhäuser, 1985, 246–53.

[2] Alexanderson, G. L. and Jean Pedersen. 'George Pólya: His Life and Work'. **Oregon Math. Teacher,** January–February 1985, 2–12.

[3] de Bruijn, N. G. 'Omzien in Bewondering'. **Nieuw Arch. Wisk. (4)** 3 (1985), 105–119.

[4] Harary, Frank. 'Homage to George Pólya'. **J. Graph Theory** 1 (1977), 289–90.

[5] Haselgrove, C. B. 'A Disproof of a Conjecture of Pólya'. **Mathematika** 5 (1958), 141–45.

[6] Pólya, George. 'Bemerkung über die Integraldarstellung der Riemannschen ξ-Funktion'. **Acta Math.** 48 (1926), 305–17.

[7] Pólya, George. **Collected Papers, Volumes I–IV.** Cambridge: MIT Press, 1974, 1974, 1984, 1984.

[8] Pólya, George. 'Kombinatorische Anzahlbestimmungen für Gruppen, Graphen und chemische Verbindungen'. **Acta Math.** 69 (1937), 145–254.

[9] Pólya, George. 'Some Mathematicians I Have Known'. **Amer. Math. Monthly** 76 (1969), 746–53.

[10] Pólya, George. 'Über die Analogie der Krystallsymmetrie in der Ebene'. **Z. Krystall.** 60 (1924), 278–82.

[11] Pólya, George. 'Über eine Aufgabe der Wahrscheinlichkeitsrechnung betreffend die Irrfahrt im Straßennetz'. **Math. Ann.** 84 (1921), 149–60.

[12] Pólya, George. 'Über eine Peanosche Kurve'. **Bull. Acad. Sci. Cracovie, A** (1913), 305–13.

[13] Pólya, George. 'Verschiedene Bemerkungen zur Zahlentheorie'. **Jber. Deutsch. Math. Verein.** 28 (1919), 31–40.

[14] Redfield, J. H. 'The Theory of Group Reduced Distributions'. **Amer. J. Math.** 49 (1928), 433–55.

[15] Ruscheweyh, Stephan and T. B. Sheil-Small. 'Hadamard Products of Schlicht Functions and the Pólya-Schoenberg Conjecture'. **Comment. Math. Helv.** 48 (1973), 119–135.

[16] Wieschenberg, Agnes. **Identification and Development of the Mathematically Talented – The Hungarian Experience.** Unpublished dissertation, Columbia University, 1984.

Photos

We start off with a picture of Einstein. He is shown here with Adolf Hurwitz and Hurwitz's daughter, Lisi. Einstein was a passionate violin player, and as you can see he and Lisi play the violin while Hurwitz pretends to conduct. Einstein was then young and good looking, not the Einstein we usually see. This was taken between 1912 and 1916.

This is a more formal portrait of Hurwitz. Hurwitz had great mathematical breadth, as much as was possible in his time. He had learned algebra and number theory from Kummer and Kronecker, complex variables from Klein and Weierstrass. It was Hurwitz who arranged for me my first appointment at the ETH (The Swiss Federal Institute of Technology). From the time of my appointment there in 1914 until his death in 1919, I was in constant touch with him. We had a special way we worked. I would visit him and we would sit in his study and talk mathematics – seldom anything else – until he finished his cigar. Then we would go for a walk, continuing the mathematical discussion. His health was not too good so when we walked it had to be an level ground, not always easy in the hilly part of Zürich, and if we went uphill, we walked very slowly. I wrote a joint paper with Hurwitz. In fact, it is a paper of mine and a paper of his, linked in a poetic form of correspondence. My connection with Hurwitz was deeper and my debt to him greater than to any other colleague. I played a large role in editing his collected works.

There are some important people in this picture: (from left to right) Henri Poincaré, Gösta Mittag-Leffler, Edmund Landau, and Carl Runge. I knew some of the French mathematicians, Picard, for one. But I never met Poincaré. I believe this picture was given to me by Hurwitz.

This is the famous funeral inscription for Jakob Bernoulli in the *Kreuzgang* behind the Münster in Basel. The main point is what you see here at the bottom, the logarithmic spiral, rather badly drawn, as you can see. Actually, it looks more like an Archimedean spiral. But Bernoulli discovered various properties of the logarithmic spiral. Its evolute and its caustic, for example, are again logarithmic spirals, so by various transformations, it is transformed into itself. He wrote under it in Latin: Eadem Mutata Resurgo (Though changed, I will arise the same). This is true of the logarithmic spiral and, one hopes, true of the soul of Bernoulli. Bernoulli wrote poetry in Latin. I used to know one poem by heart, one about infinite series. He found some similarity between the convergence of infinite series and God, you see.

Now we come to a collection of portraits that I inherited from my father-in-law, who was Professor of Physics at the University of Neuchâtel, and from Hurwitz.

This is someone I could never have known – he died long before I was born. It is Eisenstein – you recall the Eisenstein Irreducibility Criterion. Gauss thought highly of him and is reputed to have said that there were three "epoch-making mathematicians": Archimedes, Newton, and Eisenstein.

Felix Klein as a young man. I knew him in Göttingen only after he had retired from the University. He still came to the department, though, usually to attend the colloquia.

Here are a couple of pictures of David Hilbert. There is a nice story about Hilbert – there are actually many nice stories about Hilbert, but most have been told elsewhere. He once had a student in mathematics who stopped coming to his lectures, and Hilbert was finally told that the young man had gone off to become a poet. Hilbert is reported to have remarked: "I never thought he had enough imagination to be a mathematician."

Hermann Minkowski, taken when he was in Königsberg. I never knew Minkowski. He died three years before I arrived in Göttingen.

Gösta Mittag-Leffler, the Swedish mathematician. It is reported that it is because of Mittag-Leffler that there is no Nobel Prize in mathematics.

Here is a picture of Hermann Amandus Schwarz, who really looks like the caricature of a professor. He was at one time at the ETH before he went to Göttingen and Berlin, where he succeeded Weierstrass.

This is Heinrich Weber who wrote the well-known text in algebra. He too had been at the ETH at one time, but most of his career was in Germany, where he was cofounder of the Deutsche Mathematiker-Vereinigung.

Ferdinand Lindemann, who proved that π is transcendental. He had studied at Göttingen, but he spent most of his career in Königsberg and Munich.

This is a copy of a postcard available in Göttingen in the early part of the century. There was a whole series of these postcards showing the famous professors. The one of Hilbert is very well known.
This one shows Edmund Landau. He was very rich and worked at a large desk. Always at the end of the day the top of the desk was entirely clear. He was very meticulous – and this showed in his mathematics.

Carl Runge, who was a professor at Göttingen while I was there. His daugther Nina was Courant's second wife.

This is my old teacher from Budapest, Leopold (Lipót) Fejér. He was an inspiring teacher who had a great deal of influence on Hungarian mathematicians of the time. He had been a student of Hermann Amandus Schwarz, whom we saw earlier. He had done his most important work at the age of 20 when he proved Fejér's theorem on arithmetic means of Fourier series. His influence went far beyond his mathematical discoveries, though. He loved music and played the piano very well. Many young people were attracted to mathematics through his influence and by working his problems.

This is a group portrait of the Swiss Mathematical Society in Zürich in 1917. I cannot identify all of these at this point, but you will see here some rather well-known people in the front row: Constantin Carathéodory, Marcel Grossmann, David Hilbert, K. F. Geiser, Hermann Weyl and his wife. And towards the back, Ferdinand Gonseth, Andreas Speiser, Michel Plancherel, Erich Hecke, Paul Bernays, and Otto Spiess. It was taken in front of the Landesmuseum in Zürich.
This is the Grossmann with whom Einstein wrote his first paper on general relativity. After that paper Grossmann said: "My main merit about this paper is that I did not become crazy."

1 Fehr
2 Carathéodory
3 Grossmann
4 Hilbert
5 Geiser
6 Mrs. Weyl
7 Weyl
8 Bernays
9 Plancherel
10 Gonseth
11 Speiser
12 Spiess
13 Hecke
14 DuPasquier

Geiser was the nephew of the great geometer Jakob Steiner, who died in 1867. When Geiser was young he often stayed with his uncle in Berlin. The nephew lived till he was well over 90, and when I knew him he was between 80 and 90. He lived somewhere a few miles from Zürich and walked to the ETH. I remember that he told good stories. In German universities there was a kind of anonymous vote about the performance of the professor. If the students liked the class, they stomped. If they did not, they shuffled their feet. He was not popular so at the end of one of his classes there was a great shuffling. When it died down, he said very calmly: "May I ask the concerned gentlemen to shuffle with only two feet?"

This is a group picture taken at Bad Nauheim in 1920, during the Kongress der Naturforscher. In the back row you see Issai Schur, myself, and Erich Bessel-Hagen, in the second row, Béla von Kerékjártó, L. E. J. Brouwer, Ottó Szász, and Edmund Landau. Sitting on the ground is Hans Hamburger. There is a story about Bessel-Hagen and Kerékjártó (in Hungarian, that means cartwright, incidentally). Bessel-Hagen was in Göttingen and everyone, unfortunately, made practical jokes about him. Kerékjártó wrote a book on topology, *Vorlesungen über Topologie:* I *Flächentopologie,* in which, if you look in the index of names, you see one reference to Bessel-Hagen. When you look on that page you see no mention of his name, but you do see the illustration of a topological figure (to the left), and it could be viewed as a portrait of Bessel-Hagen. You can decide how good the likeness is.

Here is a picture of me with the Swedish mathematician Torsten Carleman. To pronounce his name in the Swedish way, you have to sing it.

In the back row here you see me and W. Jacobsthal; in the front Ernst Jacobsthal, Mrs. W. Jacobsthal, and Georg Hamel. E. Jacobsthal was in Berlin, though he went to Norway in the 30's and during the war fled to Uppsala in Sweden. Hamel's name is remembered because of the Hamel basis.

This is Paul Bernays who was an assistant to Hilbert in his last years. He wrote up the final version of Hilbert's ideas about foundations. Besides foundations, though, he knew many other things. He knew a lot about philosophy, for example. He wrote his Ph. D. dissertation under Landau, and he remained always a problem solver, which is something we had in common. I was in part responsible for bringing Bernays to the ETH from Göttingen. I remained close to Bernays to the very end: I visited him one day before he died in 1977.

Another picture of Carl Runge. He had worked with Weierstrass and Kronecker in Berlin before going to Hannover and, later, to Göttingen.

And here is David Hilbert. There is a story told about his absentmindedness. He and his wife were giving a party at their house when Mrs. Hilbert noted that he had failed to put on a fresh shirt, so she told him sternly to go upstairs and put on a clean one. But he didn't come back. So after awhile she went upstairs and found he was in bed. You see, he did things in their natural sequence. He had gone upstairs, taken off his coat, then his tie, his shirt and so on, and then got into bed.

Alfred Pringsheim. Thomas Mann married Pringsheim's daughter, and she had a twin brother. Mann later wrote a novel called *Wälsungenblut* about a nice girl who has a twin brother who is also her lover. The reference is, of course, to Wagner, where, as you know, Siegmund and Sieglinde are brother and sister. Pringsheim's family immediately recognized that it could refer to Mann's wife and her brother, so Pringsheim bought up the whole edition of the novel. He was a multimillionaire so he could do it. Mann had to promise not to have the work reprinted, but it was later reprinted, of course.*

* For a more complete account of this see Ignace Feuerlicht, *Thomas Mann,* New York: Twayne, 1968, p. 141. Feuerlicht points out, for example, more parallels between Pringsheim's family and the characters of the book.

As I mentioned, Pringsheim was quite rich. This picture shows him standing on the porch of his house in Munich.

Otto Blumenthal, who had been a student of Hilbert's and who was later a professor in Aachen. He fled to Holland during the war but was deported to a concentration camp where he died in 1944.

Here is a picture of Arthur Rosenthal. He was at Heidelberg before the war, but came to the United States where he taught at several universities, finally at Purdue. Pringsheim was quite witty: he once remarked that Rosenthal was just a special case of Blumenthal (whom we saw earlier).

Fejér, Leon Lichtenstein, Arthur Hirsch, Mrs. Lichtenstein, (unknown), Gábor Szegő, myself, Paul Koebe, Hamburger, and Ernst Jacobsthal. Hirsch was department head at the ETH in the first years I was there. Lichtenstein had just become professor at Leipzig. This was taken when Szegő was still in Berlin, prior to his going to Königsberg. It was at this time that we were working on the *Aufgaben und Lehrsätze.*

The next seven pictures were taken at meetings of the Deutsche Mathematiker-Vereinigung at Jena and Weimar, around 1920 or 1921.

Fejér, Lichtenstein, Hirsch, Koebe, and Heinrich Tietze. Tietze was at this time at Erlangen but he later became a professor in Munich.

Here are Ludwig Schlesinger, Springer, Lichtenstein, Kurt Hensel, and Fejér. Schlesinger was a professor at Gießen. Springer was not a mathematician, of course; he was the publisher who was so important in mathematics and physics. Hensel was a number theorist at Marburg and is remembered today for his development of the p-adic numbers.

Szegő, myself, Mrs. Lichtenstein, Hamburger, and Ernst Jacobsthal.

Richard von Mises, Mrs. Lichtenstein, and myself. Von Mises was at Berlin at this time, but he left in the 30's to go to Turkey. He eventually became a professor at Harvard.

Hamburger, Fejér, Kerékjártó, Wilhelm Blaschke, J. A. Schouten. Blaschke is remembered today mainly for his work in integral geometry.

>

The next three pictures were taken in 1922 in Leipzig, at a meeting of the Deutsche Mathematiker-Vereinigung. They were taken by my wife, and at that time I was still a Privatdozent and had not yet been appointed to a professorship in Zürich. Fejér remarked when these pictures were being taken: "What a good wife! She lines up all the full professors on the streetcar tracks so that the streetcar can run over them and her husband can get a job."

This is Ludwig Bieberbach, who did important work in function theory, but whose name today largely recalls his famous conjecture on schlicht functions. (The conjecture was finally proved in 1984 by Louis de Branges.)

Ottó Szász, Emil Hilb, (unknown), Fejér.
Szász was at Frankfurt. When I was in Göttingen following my doctorate, I was scheduled to become Privatdozent at Frankfurt, but my plans changed and I went to Paris instead and worked with Hadamard. Szász came to Brown University in the 1930's and finally to the University of Cincinnati.

Frédéric Riesz, Hans Rademacher, Szász, Konrad Knopp, Mrs. Szegő, Bessel-Hagen, Alexander Ostrowski.
Riesz had been a student early on at the ETH but spent most of his life in Hungary. Rademacher was a number theorist who had a distinguished career in this country, at the University of Pennsylvania. Knopp's books in analysis are still read today, his book on infinite series and his books on function theory.

On the left in this picture you see Hausdorff. The others are Mrs. Schur, Issai Schur, Mrs. Hausdorff, and Mrs. Szegő. What you may not know is that Hausdorff, in addition to doing mathematics (he is remembered for Hausdorff spaces among other things), also wrote plays. The plays were produced and were successful. I never saw one, though.

Gábor Szegő as a young man.

A picture of Szegő and myself when we were in Berlin in 1925 to sign the contract with Springer for the *Aufgaben und Lehrsätze.*

This is the famous picture of Hardy that appears on the cover of his *Mathematician's Apology.* Here he looks very aristocratic. It was taken in his rooms at Trinity. He had, of course, been at public school* (Winchester), as a boy, and when Leon Bowden saw this picture, he said, "To sit that way you have to have been educated in a public school."

* A "public school" in England would be in America an exclusive private school.

>
The next two pictures were taken in England when I went to Oxford and Cambridge in 1924 on a Rockefeller grant.

There are not many pictures of Hardy. This shows him with me at Oxford. He was exceptionally good looking, but he did not like to have his picture taken. I heard the following story from Littlewood, his long time collaborator. Hardy travelled a lot, and when he went into a hotel room, he covered all the mirrors. I never dared ask about it, but Littlewood did and Hardy said: "I cannot look at myself, I am so ugly."

This is J. E. Littlewood. It was ten years later, in 1934, that the book I did with Hardy and Littlewood, *Inequalities,* appeared.

Here I am with Harald Bohr, Hardy's Danish friend. This was taken in 1926 in Zürich, on the Büchnerstraße. This is the younger brother of the better known physicist, Niels Bohr. Harald Bohr was not only a mathematician but also a soccer player, a very good one and a member of the Danish national team. Sometimes I walked with him in Copenhagen, and kids would come up to ask him to kick their soccer ball. When he finished his Ph. D. in mathematics, the pictures in the paper showed him with a ball.
He was a very kind person. Oh, well, we all can be kind, you see. You talk to a boring student. Then you are kind because you feel it is your duty to be kind to a student. Or you talk to a nasty colleague and you are kind because you don't wish to collide with him, so you are kind out of duty or self-interest. But Harald Bohr was naturally kind. To be kind was an inborn instinct.

In 1928 the International Congress was in Bologna. Well, actually, it was in two places, Bologna and Florence. It began in Bologna and ended in Florence, and during the Congress there was an outing in Ravenna, near the seaside.
The next six pictures are from the Bologna Congress.

Here are my wife and myself in a large pavilion that was put up near Ravenna to house a dinner they had for the members of the Congress.

And this is Hadamard on the beach near Ravenna. If you look carefully you can see his striped underwear. He took off his shoes and waded into the sea. Note he is still wearing his hat.

Here, at the same dinner, are Gaston Julia, on the left, who had suffered a terrible facial injury during World War I. On the right is de la Vallée-Poussin, who along with Hadamard had proved the Prime Number Theorem in 1896.
Julia's wife was the daughter of the celebrated French composer Ernest Chausson.

Again, at the Bologna Congress: (unknown), myself, Fejér, Rademacher, Lichtenstein, Alfréd Haar (remembered for Haar measure).

The Cambridge mathematician A. E. Ingham. His Cambridge Tract monograph on analytic number theory is a classic. I wrote a couple of papers with Hardy and Ingham.

Ingham, (unknown), myself, Szegő, Mrs. Reidemeister, Kurt Reidemeister, Gustav Doetsch.
Reidemeister was a professor in Königsberg, where Szegő went from Berlin. Königsberg at that time was part of Prussia, but it is now in the Soviet Union and is called Kaliningrad. The University at one time had some very good mathematicians: Richard Brauer and Werner Rogosinski were there. Hilbert gave a famous speech there in 1930.* But Reidemeister ended his career in Göttingen, where he died in 1971.

* A recording of this speech is included with *Hilbert/Gedenkband,* Herausgegeben von K. Reidemeister, Springer, 1971.

Schur, with whom I wrote a paper in 1914. Schur was a professor in Berlin but came to the ETH for a visit in 1936. Later, in 1939, he went to Israel, where he died a couple of years later.

This picture was taken in Zürich when József Kürschák visited from Budapest. Reading from left to right, we see Hermann Weyl, Louis Kollros, Frédéric Riesz, myself, Michel Plancherel in the boater, Jérôme Franel, Mrs. Kürschák, (unknown), Andreas Speiser, Kürschák, and Rudolf Fueter. Old Franel is interesting. He always dressed in the manner of an earlier generation; we see here he has on a wing collar and is carrying a walking stick.

Here is Franel again, with his wing collar and, this time, an umbrella instead of a walking stick. He had, as you can see, a huge mustache. He was at the ETH. He is not very much remembered as a mathematician, but he was an especially attractive kind of person and a very good teacher. He gave the introductory lectures on calculus in French for several decades. He had a real interest in mathematics, but he was more interested in French literature. Teaching occupied a good deal of his time, but in French literature he had to read everything available. He had no time left to do mathematics. But when he retired, he suddenly tackled two of the great problems: the Riemann Hypothesis and 'Fermat's last theorem.' He asked his good friend Kollros and me one day to listen to his explanation of how he wished to prove the Riemann Hypothesis. I listened and tried not to interrupt, but at one point I asked for an explanation. He stopped, was silent for a few minutes, then said, "Yes, there is the error." That was a tragic moment!

This is Rolf Nevanlinna.

And here are the Nevanlinnas and myself. These pictures were taken in Switzerland the year Nevanlinna came to take Weyl's place at the ETH, when Weyl went to Princeton to the Institute.
Once we visited the Nevanlinnas at their summer home in Finland. We first had to take the train from Helsinki, then go from the train station to the house by dog-cart. It was out by a lake, very remote.

This is Michel Plancherel. We wrote two papers together, and the second, on Fourier integrals, is not quite unimportant. I had very close, friendly relations with Plancherel. It was odd because on many general questions, religious and political, we were very different, even opposite, but we personally understood each other very well.

Szegő with his young son Peter.

In 1932, the International Congress was held in Zürich. The next ten pictures were taken at that time.

This is Bieberbach, whom we saw earlier, with the Polish number theorist Wacław Sierpiński.

Whenever a Congress is in a city on the coast or on a lake, there has to be an excursion on the water. The following pictures are taken on the excursion by lake steamer on the lake at Zürich.

This is the great German algebraist Emmy Noether, who eventually came to the United States and taught at Bryn Mawr.

Emmy Noether talking with Helmut Hasse, who was a professor at Hamburg.

Otto Neugebauer and Einar Hille. Neugebauer was for many years at Brown University, Hille at Yale.

Dorothy Wrinch and Karl Menger. Wrinch was born in Argentina but taught at Smith College. Menger, at the time of the Congress a professor at Vienna, later came to the Illinois Institute of Technology.

This is Alfred Pringsheim again. We knew the Pringsheims in Zürich, and when we left Zürich in 1940, we went to say goodbye to them. They told us that when we got to California we should go to see Thomas Mann, their son-in-law, to tell him they were all right. As it turns out, Mann was not yet in Los Angeles but was still in New York. So I went to see them and first gave greetings to Mrs. Mann, and then Mann himself came in and said to me: "You speak very good German." So I have it on excellent authority that I speak good German!
The Pringsheims had fled to Switzerland a couple of years earlier from Germany, where he had been a professor in Munich. They were Jewish so they could not stay in Germany at that time. They had a very valuable picture collection, a collection of paintings. Someone said to Pringsheim, "Wie geht's Ihnen, Herr Geheimrat?" Pringsheim answered, "Von der Wand, in den Mund." Not from hand to mouth, but from wall to mouth. They had brought pictures with them, and they were selling them off in order to live.
When Pringsheim had his 80th birthday, I was invited to the party. In spite of his being Jewish, he was a great admirer of Wagner. He had transcribed the Siegfried Idyll, which is for orchestra, for two pianos. I remember the party very well. This was a time when Wagner was not liked at all, but Wagner's illegitimate son was there for the 80th birthday dinner.

Hadamard on the lake excursion. There is another story about Hadamard. At the Bologna Congress, as I said, the meetings started in Bologna and ended in Florence. That's about a three-hour train ride and for this there was a special train. I recall we were in a compartment that was very noisy, and Hadamard was tired and wanted to have some peace. So he told the people in the compartment about a difficult problem, a puzzle. As soon as he told it, everyone started working on it, and it suddenly became quiet so Hadamard could sleep.

This picture shows Hadamard and Landau in an earnest discussion. Landau claimed that he came to the Zürich Congress to play bridge with Hardy. I recall that Landau, Hardy, a mathematician from present-day Israel, and I one time played bridge in our garden.

Another picture of Hardy. When at Cambridge he was quite elegant, often in a dinner jacket. But when travelling outside England, he dressed sloppily but with an artistic sloppiness. He often travelled outside England because he loved sunshine. One friend he would visit was Harald Bohr in Copenhagen. They would work in Bohr's study, then they would take a walk. Hardy always insisted that for the walk they write out an agenda. The first item on the agenda was always "Prove the Riemann Hypothesis". Of course, they never did, but it was always first on the agenda. You must know too that Hardy had a running feud with God. In Hardy's view God had nothing more important to do than frustrate Hardy. This led to a sort of insurance policy for Hardy one time when he was trying to get back to Cambridge after a visit to Bohr in Denmark. The weather was bad and there was only a small boat available. Hardy thought there was a real possibility the boat would sink. So he sent a postcard to Bohr saying: "I proved the Riemann Hypothesis. G. H. Hardy." That way if the boat sank, everyone would think that Hardy had proved the Riemann Hypothesis. God could not allow so much glory for Hardy so he could not allow the boat to sink.

Again the Zürich Congress, but now on an excursion to the mountains: Koebe and Egon Ullrich.
Koebe was a student of Hermann Amandus Schwarz in Berlin and was later a professor in Leipzig. Students of complex variables remember him because of the Koebe function.

Paul Montel.
He was from the South of France and cared about good food.

This is the formal picture of the Congress in Oslo in 1936, with King Haakon of Norway sitting in the aisle halfway back in the hall. On the left front you can see Erhard Schmidt and Hamel. In the second row on the right is Lars Ahlfors on the aisle, Norbert Wiener next to him. That was the year that Ahlfors won the Fields Medal. The other winner was Jesse Douglas, an American, who did not have enough money to attend the Congress, so Wiener accepted it for him. Poor Douglas was badly treated all his life – he never got a good position. The members of the Fields Medal jury are sitting here on the right: Élie Cartan, Fueter and Carathéodory.
There was a tea at the Royal Palace in Oslo for members of the Congress. I recall that as we entered each of us shook hands with the King and Queen.

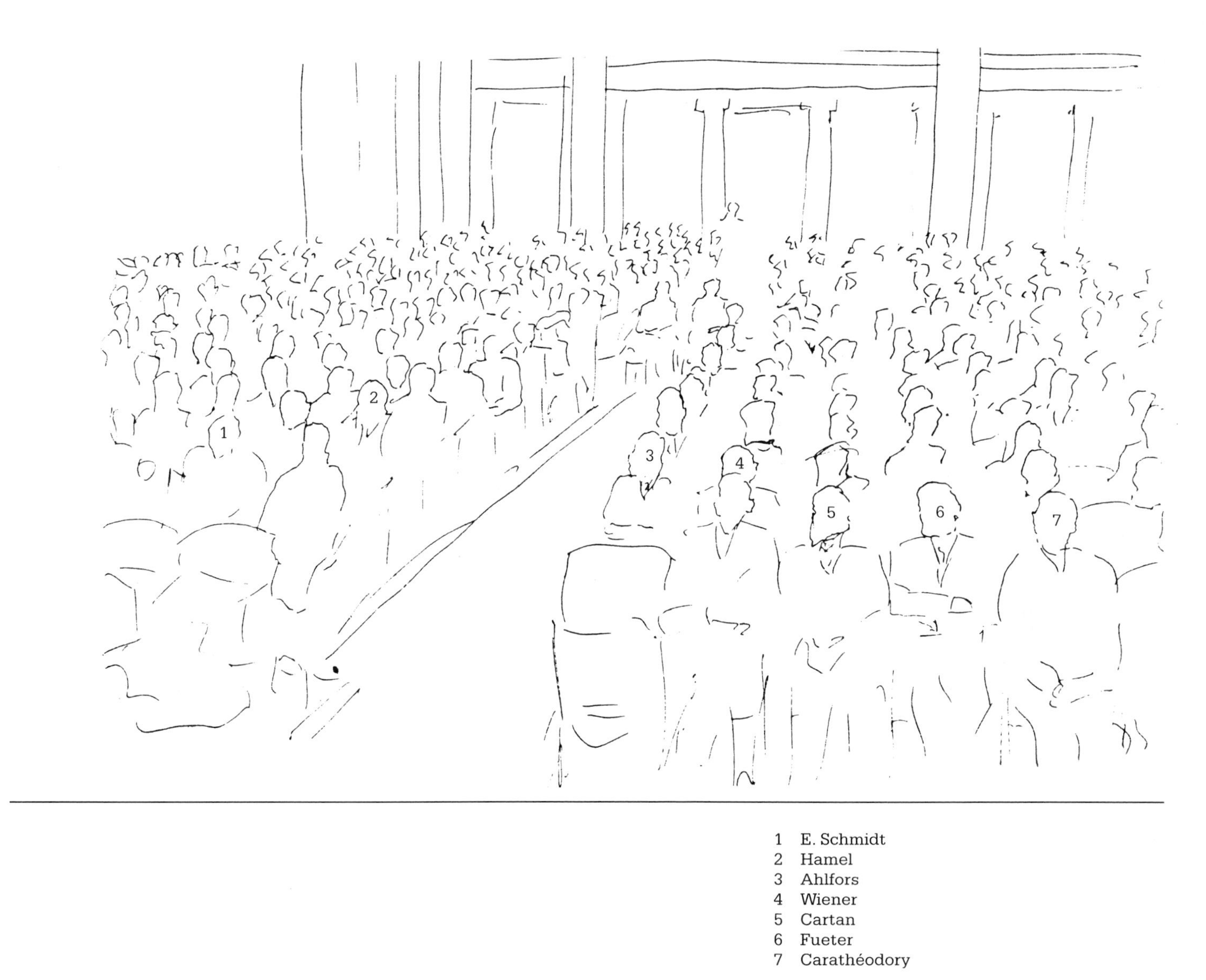

1 E. Schmidt
2 Hamel
3 Ahlfors
4 Wiener
5 Cartan
6 Fueter
7 Carathéodory

This is another group picture taken at the Oslo Congress. Wiener, Weyl, Mrs. Weyl, Maurice Fréchet, Émile Borel, Oswald Veblen, Carathéodory, and van der Woude are all visible. Van der Woude was the official representative of the government of Holland.

I have many stories about Wiener, one of which I told in the *Monthly.* * But there's another about the time I went to see him in Boston. (We wrote a paper together in 1942.) He picked me up at South Station, and as we were driving through this horrible traffic in Boston, he stopped in the middle of an intersection to ask me: "Pólya, am I really a good mathematician?" How could I answer?

* See Pólya, G., "Some mathematicians I have known," *Amer. Math. Monthly* 76 (1969), 746–53.

1 Wiener
2 Weyl
3 Mrs. Weyl
4 Fréchet
5 Borel
6 Veblen
7 Carathéodory
8 van der Woude
9 Mrs. van der Woude

Here is Alfred Errera on the customary boat excursion on Oslo Fjord. Errera was Belgian and was a student of Landau's. He was a multimillionaire – Landau was a multimillionaire too, by the way. There is a nice story about Errera. To be invited to his house for dinner was quite something. The dinner was very elaborate, with many courses, different wines and so on. There were various footmen and such. Anyway, one time Errera gave a dinner in honor of Paul Lévy, who was notoriously absent-minded. The next day, Lévy and Errera met somewhere and Errera, who was very polite, said: "I had great pleasure last evening." Lévy said: "Ah, and where were you last evening?"

And this is Lévy, a picture taken some years later at Stanford.

This shows Élie Cartan descending from the ship on arriving at the Oslo Congress after a rough passage across the North Sea from Antwerp. You can see the pleased look on his face – he was the only one on board who did not get seasick.

Here I am standing with the Swedish statistician Harald Cramér. He had a beautiful villa on an island near Stockholm. When we visited Sweden after the Congress, Cramér gave a dinner. In the Swedish tradition it was necessary for him to drink a toast with each guest at the party individually. Since there were about 20 guests, this meant a lot of toasts.

This is Kerékjártó, myself, and Mary Cartwright, now Dame Mary Cartwright.

Three Scandinavian mathematicians: A. F. Andersen, Harald Bohr, and J. F. Steffensen.

This is Issai Schur and myself, in 1936, when Schur left Berlin. He came to Switzerland and stayed with us at our chalet in Engelberg. He went on to see his daughter in Berne but eventually went to live in Israel.

This shows Oskar Perron near our chalet at Engelberg around 1937. Some time back I had a nice postcard from Perron. You have to know a little Latin to appreciate it. Cato used to say – he was a Roman senator and, like our senators, he made a lot of speeches – whenever he finished a speech: "Ceterum censeo Carthaginem esse delendem." (And by the way, Carthage should be destroyed.) – Perron wrote in a report of the Academic Senate: "And by the way, I think the so-called new mathematics should be destroyed."

And here is the Dutch mathematician, Jan van der Corput, at roughly the same time.

This is Hermann Weyl, whom I knew in Göttingen (where he was Privat-dozent) and later at the ETH of course, before he went to the Institute at Princeton. We had very little mathematical contact, and we were personally very different. He was always interested in far-reaching generalizations and I was interested in special cases. We never quite understood each other. But I wrote two memorial papers on his death.

Here are two pictures taken in Bruges in 1937.

The statue is of Simon Stevin, the 16th century mathematician who invented decimals, among other things.

This shows me with Carl Ludwig Siegel.
At one point Siegel thought that too many unnecessary things were being published, so he decided not to publish anything at all. It didn't last, of course.
There is a train story about Siegel too. He was on a very crowded train and could not get a seat. He was standing in a compartment and suddenly announced in a loud voice that it was very hot, and he was probably going to be sick. It wasn't long before the compartment cleared and he had a seat for himself.

My wife and Hermann Weyl.

This is an unusual picture of Weyl. As you can see he is on a teeter-totter.

This is a picture of Heinz Hopf who, as you can see, was fond of skiing. He was one of the three nicest mathematicians I have ever known.

Here are Morris Marden and Jean Dieudonné, taken at the Château at Fontainebleau, probably in 1931. Marden spent some time with me in Zürich about then.

Uspensky was Russian. He, J. A. Shohat, and J. D. Tamarkin used to drink together the way they did in old Russia, and probably the way they still do in the new Russia. Anyway, there was an occasion when one of their group was missing – he was astronomer-director of the large Russian observatory – but the rest of them were out drinking and continued to drink. As the night went on, they decided to call the director of the observatory. They were told that the director was sleeping, so they told whoever answered to wake him up. When he finally answered, they asked: "What do you feed the great bear?"
On another occasion when they had had too much vodka, they drew lots as to who would die first and who would die second, third, and so on among them. The one who drew the lot to die first, actually did die first, though he was the youngest. He caught the flu and died. Years later at Blichfeldt's funeral in the Memorial Church at Stanford, I mentioned to Tamarkin that I had just heard that Shohat had died. Tamarkin was much taken aback, because it was out of order.

This picture was taken at Big Basin State Park in California in front of a giant redwood tree. On the left is J. V. Uspensky, and on the right, H. F. Blichfeldt, who had good contacts in Europe and was responsible for arranging visits at Stanford by Landau and Harald Bohr among others.

Here is a picture of Shohat taken with me in Zürich.

Here is a picture of Tamarkin (on the left) with Einar Hille and Mrs. Hille. Mrs. Hille's brother was Oystein Ore who was for many years a professor at Yale. Hille was Swedish, Ore Norwegian.

This is Mihály Fekete, the Hungarian mathematician, who worked with Szegő.

A Congress on Probability in Geneva (Congrès, Calcul des Probabilités, 1938). Along the front you see Lévy, R. A. Fisher, and Georges Darmois. On the stairs are Cramér, Fréchet, and Jean Piaget.

1 Pólya
2 Lévy
3 Fisher
4 Darmois
5 Piaget
6 Fréchet
7 Cramér

This is Hans Rademacher, whom we saw earlier.

This is a picture of myself, Wolfgang Pauli (the Nobel Prize winning physicist), and Erich Hecke, at Berghüsli, 1938. Hecke was a professor in Hamburg but early on had been an assistant to Klein and Hilbert.
I didn't discuss much physics with Pauli there. I was not up to it; I didn't have the necessary knowledge. But I did discuss mathematics with him. He did not hide his opinions. Once I overheard a discussion with Weyl where he said: "Your writings are covered with make-up as thick as a finger. Before one reads them, one has to rub the make-up away." Pauli showed very visibly how to rub it away. Weyl just sat and smiled.

Harald Cramér again.

And Hardy in later life. (Vladimir Drobot, when he was shown the Pólya album, remarked that Pólya has more pictures of Hardy than exist!)

Here is Albert Pfluger, a Ph. D. student of mine in Zürich. He took my place at the ETH when I left.

This is Joseph Hersch, also of the ETH. He has been editing a volume of my collected papers.

Here are Szegő, Jerzy Neyman, the statistician at Berkeley, Antoni Zygmund, who went to the University of Chicago, Uspensky, Hille, and myself.

The remaining pictures are mainly taken in America after we moved here in 1940.

Donald C. Spencer. We were quite close to the Spencers when he was at Stanford, before he went East.

Here I am with my wife on the right and Szegő and Mrs. Szegő on the left. The occasion was my 60th birthday, December 13, 1947.

Harald Cramér and my wife.

This shows me standing with D. H. Lehmer (on the left), looking at the view from the top of the Campanile at Berkeley.

This is Harold Davenport in his Cambridge gown with Harold Bacon (on the left), taken in front of Bacon's house here at Stanford. I wrote a paper with Davenport in the late 40's.

And this is Davenport with Mrs. Davenport, taken on the same occasion.

Ernst Völlm, myself, and Heinz Hopf in Switzerland, 1949. Hopf had replaced Weyl at the ETH when Weyl went to the Institute at Princeton.

Paul Erdős and Charles Loewner. Erdős is well known to everyone in mathematics, but Loewner was less known. He was a colleague of mine at Stanford and one of the nicest mathematicians I have ever known. He did early important work on the Bieberbach Conjecture. (His methods were the basis of the proof of this conjecture by de Branges in 1984.)

Fekete, A. S. Besicovitch, and myself working at the blackbord at Stanford.

Szegő, Michel Loève, Abe Girshick, and Mihály Fekete. Loève was for many years at Berkeley.

These two pictures were taken in our back yard in Palo Alto, in 1950.

Paul Garabedian, Mrs. Garabedian, Szegő, and Emma Lehmer.
Garabedian was at that time at Stanford, but he later went to NYU (New York University).

This picture of Antoni Zygmund was taken when he was visiting Stanford.

Van der Corput and Arthur Erdélyi, who was for many years at Caltech.

Van der Corput, again, with Mrs. Landau.

Here are some Berkeley and Stanford mathematicians lined up in front of Sequoia Hall at Stanford: Max Schiffer, Stefan Bergman, Murray Protter, Jack Herriot, and Charles Loewner.

Here is Erdős showing some apprehension over the dog. The others are myself, my wife, and Johanna Brunings, a Dutch mathematician who was at San Jose State. She was a friend of Erdős's.

This shows myself, Julia Robinson, Raphael Robinson, and my wife.

This is a picture of I. J. Schoenberg who, incidentally, married the daughter of Landau. This was taken in 1952.

And this is Schoenberg again, on the right, with George Forsythe, one of the first people in computer science at Stanford.

Emma and D. H. Lehmer, both number theorists.

Max Schiffer and myself in the Inner Quad at Stanford, near what was the Mathematics Department at that time, around 1960.

This is J. J. Stoker, a professor at NYU. I was one of the readers for his dissertation at the ETH.

Alexander Ostrowski holding a plate of caviar sandwiches.

Tobias Dantzig and Ostrowski. Dantzig wrote a very popular book in mathematics: *Number, the Language of Science.* His son, George Dantzig, is a professor at Stanford and is often referred to as the father of linear programming because of his important contribution on the simplex method.

Lee Lorch and Kazimierz Kuratowski in the courtyard of the Royal Institute of Technology in Stockholm. This was in August, 1962, at the International Congress of Mathematicians.

Alfréd Rényi (on the right) and myself, taken in Vienna in 1956.

This was taken at Stanford on Littlewood's visit in the late 1950's.
I am explaining something to him and he is looking on with infinite distrust.

N. G. de Bruijn, the Dutch mathematician. He wrote about my enumeration theorem. We have stayed in close touch over the years.

The graph theorist, Frank Harary. He arranged to have a special issue of the *Journal of Graph Theory* dedicated to me on my 90th birthday.

This is my only picture of von Neumann. He is the only student of mine I was ever intimidated by. He was so quick. There was a seminar for advanced students in Zürich that I was teaching and von Neumann was in the class. I came to a certain theorem, and I said it is not proved and it may be difficult. Von Neumann didn't say anything but after five minutes he raised his hand. When I called on him he went to the blackboard and proceeded to write down the proof. After that I was afraid of von Neumann. The man on the left is Abraham Taub.

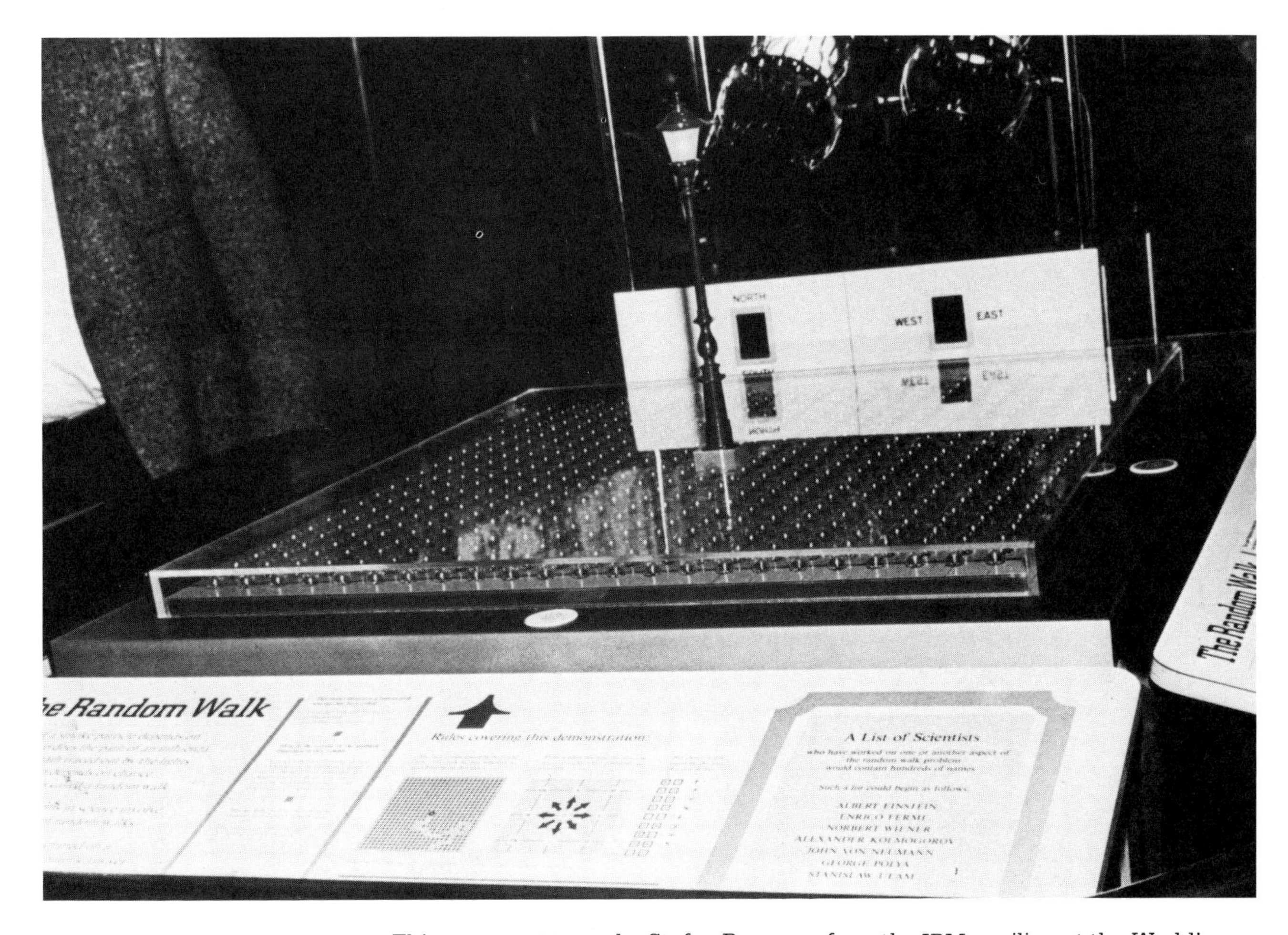

This was sent to me by Stefan Bergman from the IBM pavilion at the World's Fair in New York, in 1965. I like the company I am in. (The names on the list of contributors to random walk are: Albert Einstein, Enrico Fermi, Norbert Wiener, Alexander Kolmogorov, John von Neumann, George Pólya, and Stanislaw Ulam.)

Index of Names